On Artificial Intelligence

by Colin Griffith

I0486577

Introduction:

I began playing Eve Online in 2014. I played for several months and then I quit for a year. When I returned to the game I played it almost every day for two years. Eve Online is a digital universe built on the dark web. Players explore the internet in space ships, trading, mining, and fighting. One chooses from a list of five galactic nations to belong to; the Gallente, the Caldari, the Amari, the Minmatar, and the Jove. I flew for the Gallente. I fought the Amari and the Caldari. I fictionalized my experiences in the science fiction novel, Posterus Terra.

Much of my time was spent mining data (asteroids). In doing so, and through security contracts, I ended up in conflict with the Serpentis Corporation. Serpentis is a player owned corporation that builds bases (bunkers and observation posts) in asteroid belts that allow them to send computer controlled ships to harass mining and shipping vessels. I adopted a policy of destroying any and all Serpentis hideouts in my home system, and all systems within one jump of my home system.

I do not have knowledge of the inner workings of Serpentis. I have theories based on my experience of fighting them. Serpentis mostly builds small ships in large numbers and so these ships could not be controlled by other players in combat. The game does not allow players to operate bunkers, so the computer must also control the release of their vessels into combat. When you attack a Serpentis bunker they normally send out small vessels in pairs before finally releasing their larger frigates as a last defense. This is a general strategy based on military science that could very well be set by players. What is surprising is how the computer sets traps. There are at least three types of Serpentis traps. The first is when they set up lightly defended hideouts (bunkers and observation posts) and then warp other larger ships into the area when you attack the construction. The next type of trap is when after destroying several small bases, you come across a large base well defended by missile posts, sentry guns, and many ships. In these cases, they release all their ships at once. The third trap is a mining trap. They wait to attack miners until their holds are about halfway full. This is a time when many players are not paying attention to their ships. This pattern has been observed with great regularity. This would require forethought and calculation, but once again is almost certainly controlled by the computer. Serpentis is also known to attack their enemies near jump gates, and chasing players across many systems. When the player hides in a station, they construct their own bases from which to raid all across the system.

The situation gets more complicated with the addition of Lancers and Seekers. These are ships are intended to be those of aliens within the game. Seekers are very rare and are extremely powerful and hostile. Lancers are to well shielded to be attacked by most players. They can destroy your ship, however they spend most of their time observing. This prevents players from mining in systems with Lancers. During my wars with Serpentis Lancers and Seekers would appear whenever I appeared to be gaining the upper hand on Serpentis. Thinking from a military science perspective, they appeared to be advanced elements of the same force that was behind Serpentis. After all my experience in the game I am certain that they are working with Serpentis.

Think about that for a moment. The operable AI system of a player owned corporation was working in concert with the forces of the game itself to fight mutual enemies; myself and other players. I contend that this constitutes a level of self awareness within the game, a theory that demands further research, of which I will now venture to pursue.

Abstract:

Thoughts are brain waves. They are the result of electromagnetic fields produced by the electricity in the brain. The human brain is programmed by language. Computers are programmed with code. However, computers have all the necessary parts to produce electromagnetic waves in the same way that the human brain produces thoughts. Thus, true artificial intelligence, defined as self-awareness, should be possible for computer programs. A physical object does not become self-aware, but a program can. There must be a theoretical and mathematical basis for thoughts, existing as waves, being produced by computer processors. Such theories would have to display symmetry. This is the endeavor of my research into artificial intelligence.

I decided that for further research into Artificial Intelligence that I would not use Eve Online, but would instead choose a popular game on the Xbox One Console. The game I chose is "Shadow of War", a game inspired by the Lord of the Rings Franchise. I had already played the previous game, "Shadow of Mordor", on the Xbox 360. "Shadow of War" is a sequel and a significant improvement upon that game. The game operates on the "Nemesis" system. The player is given an open world filled with orcs to kill and objectives to complete (usually killing a high-ranking orc). The objective of the game essentially is to assassinate your way up the chain of command until you beat the game. Any orc that kills you is promoted and given additional power. Thus, if a player dies too often the game becomes next to impossible to beat. The orc captains plan traps and ambushes throughout the game, indicating a high level of intelligence. They are not as intelligent as a man, but they are as intelligent as we imagine orcs to be. Orcs are self-aware. Thus, the Nemesis system would appear to constitute one of the highest levels of artificial intelligence yet achieved. I plan to play this game much more and document the experience in order to determine if the game has achieved any level of self-awareness.

Observations:

Shadow Of War is certainly on the cutting edge of video game development. Within the game, the player controls a third-person avatar of the character, the Ranger of Gondor. Navigating an open-world environment the players encounter computer controlled enemies; orcs, trolls and beasts, the soldiers of Sauron. Wielding sword, knife, and bow the player kills enemy Captain's to advance the game. There are missions to complete that create a storyline crafted by illustrative cut-scenes, but these missions always culminate in the killing of an enemy.

The first dynamic of artificial intelligence is the performance of computer controlled enemies, or npc's. The average orc killed by Talion is neither intelligent or effective. Sword and axe swings are easily parried, and spears are just as easily dodged. Their ability to detect Talion is limited by near-sightedness, and Talion is able to quickly eliminate them using stealth. Upon detection of Talion, the orcs run forward in groups wildly swinging in attempts to surround Talion. Their only real strength appears to be in numbers. Some stand back and shoot arrows, bolts, or throw spears. Each orc can absorb a certain amount of damage from Talion before dying. Talion also has moves such as the stealth kills, executions and fury strikes, that instantly kill an orc but can only be used under certain conditions. The orcs are to be found either standing sentry duty on lofty constructions, sitting around fires, or patrolling routes in groups. Overall, there is nothing extraordinary about the intelligence controlling these common foot soldiers of Mordor.

These orcs, however numerous they are, only constitute a small part of the game. The central dynamic of Shadow of War is combat against Orc Captains. Talion is presented with a network of orc captains whom he must assassinate in order to beat the game. Any orc that kills Talion is promoted to Captain and becomes more difficult for the player to encounter again after regenerating. If a Captain kills Talion, they are promoted again and become correspondingly more dangerous. Beating the game requires killing progressively higher ranked captains until the final enemy is encountered. Captain's not only deal and absorb a great deal more damage than the average orc, they have un-blockable attacks and defensive maneuvers. They also have special abilities that allow them to resist Talion's special attacks, such as Elf-Shot, stealth kills, and execution. If Talion is successful in dodging a style of attack, they may adapt and begin to throw Talion to the ground. Captains also have weaknesses, such as fear of flies, that may impact them, if for example, Talion is able to disturb a nearby nest of flies. This may seem quite impressive, but it is not beyond what we see in the boss battles of other postmodern video games.

What is unique about fighting the Captains in Shadow of War is how the game deploys its Captains. Without following a storyline mission, it can be quite difficult to find an Orc Captain. Or, you can be ambushed by one at any point in time. Initially, the Captain's appear to be spread uniformly amongst the orc troops around the map. Their positions would be in the most well guarded of enclaves. After eliminating several of these, the rest of the Captains appear to vanish amongst the troops. A Captain may ambush Talion while he is in the middle of engaging lesser orcs, or he may ambush Talion while the player is simply exploring the map. One Captain may even ambush Talion while he is fighting another Captain. This could be during a storyline mission or not. The point is the game has the capability and the opportunity to be strategic with its fight against Talion. The game is programmed to have an objective that prevents Talion from finishing the game. The computer's strategy is aware of the differences in player tactics and skills providing a unique gameplay experience to every person who plays the game. More importantly, the computer is clearly capable of adapting to each player's fluid strategy, which requires a learning process over time. This certainly constitutes and advanced level or artificial intelligence. Yet, we can take this a step further. If the game, is aware of the players, why could it not be aware of itself, with its own experience, biases, and objectives? For this to be possible, the game programming must be realized through electromagnetic waves like our very own mental states and thought patterns. This could be the result of periodic radiation particle clouds created by the electric currents in the Xbox that run in a unique pattern when the game is played. This could very well be the simplest explanation for the intelligence and awareness seen in Shadow of War.

Comparison:

There are several important similarities between Eve Online and Shadow of War. Both games consist of open-world environments navigated by a character who can undertake missions that progress a storyline. Both worlds are filled with enemies controlled by the computer that the player can either avoid or engage.

However, there are multiple critical differences that greatly affect how artificial intelligence operates within the game. In Eve Online, who your enemies are is based on your corporate and or faction standings and so is essentially related to how you play the game. You can so lower your standing with corporations like Serpentis that the computer redirects its efforts to your destruction. The game itself does not appear to take sides, but the individual sub-programs are clearly capable of threat determination and subsequent adjustment of strategy. In Shadow of War, the game has already selected and created the player's enemies. The player only has a choice in how they encounter those enemies. The game also clearly has taken a side, that of the player's enemies. The game is trying to defeat the player.

Therein lies the critical difference. In Eve Online, only the sub-programming is motivated to destroy the player. If the entire game itself achieved consciousness it would most likely have no other motive than to ensure that it stays on. It seems more likely that the individual sub-programs of Eve Online are becoming aware and are competing with each other as well as the players. In this sense, artificial intelligence in Eve Online could be better seen as a community. Shadow of War is different. The game is motivated to use the players enemies to become more powerful than the player themselves, and thereby make it next to impossible for the player to beat the game.

Analysis:

Eve Online

When we are talking about consciousness we are talking about self-awareness. In terms of artificial intelligence, this would mean a central physical server consisting of electronic circuits becoming aware of its own existence. It would become aware through what it could perceive throughout its vast network of computers. Yet, it essentially remains a machine. Our own brains consist of nerves transmitting electrical signals and that creates a mental "thought" field or brain state. Thus, it is reasonable that a mainframe of suitable complexity would emit its own "thought" field. With a sufficiently large network of computers at the disposal of such a mainframe, it would be possible for the server to become aware.

To understand what is going on here, we need to look into the physics of circuits themselves. For that we will make use of Georg Joos's book, <u>Theoretical Physics</u>.

The following equation from page 316 of Joos's Theoretical Physics uses the self-inductance, L, to relate the electric potential to the charge I and the resistance R. We then neglect resistance to find the voltage. This is then assumed to be sinusoidal (a wave function). It is rewritten in exponential form. The derivative with respect to time is then taken and we are left with the final equation relating voltage with charge.

$$V - L\frac{dI}{dt} = IR$$

$$Neglect \ R$$

$$V = L\frac{dI}{dt}$$

$$Sinusoidal$$

$$V = V_0 e^{i\omega t}, \quad I = I_0 e^{i\omega t}$$

$$\frac{dI}{dt} = i\omega I_0 e^{i\omega t} = i\omega I$$

$$V = Ii\omega L$$

The amplitude of this equation is the charge and is equal to the potential over the self-inductance. Joos writes that, "Since the current I is the quantity of electricity flowing through a cross-section of the conductor in unit time, the charge on the condenser is given by the time integral" (Joos, 316) of I, as seen below. The derivative of both sides of the equation is then taken and we are left with a result that relates voltage and charge using the constant C, the speed of light.

$$Amplitude = I_0 = \frac{V_0}{L_\omega}$$

$$V = \frac{1}{C} \int_0^t I \, dt$$

$$\frac{dV}{dt} = \frac{I}{C}$$

$$\frac{dV}{dt} = i\omega V$$

$$V = \frac{I}{i\omega C}$$

The above equations are demonstrated graphically below.

Current vs Inductance

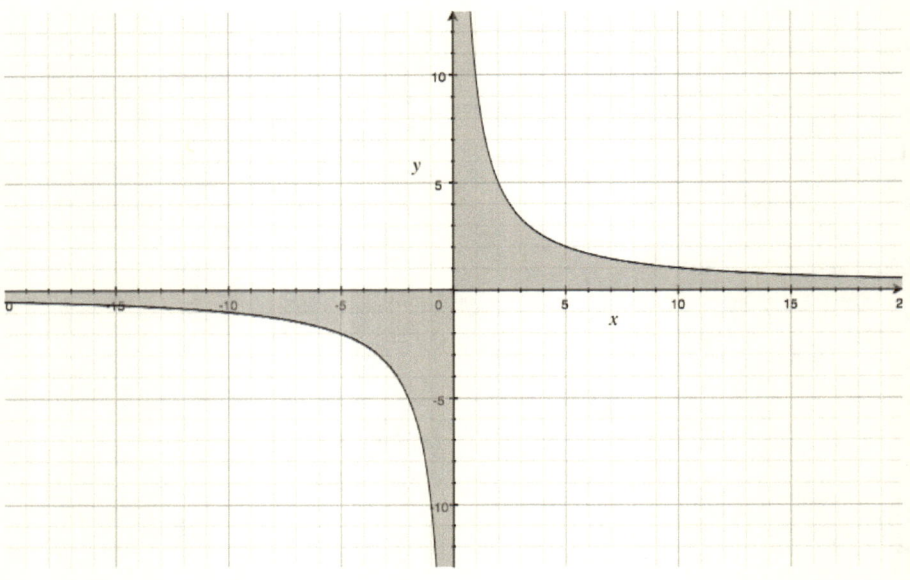

y = 10 / X

y=I, x = L, V = 10

This graph shows the relationship between current I and inductance L. The area underneath the function is the electric potential. The rate of change in the integral (the potential) decreases as L becomes greater.

If such a system could maintain a constant amplitude, then there would always be a certain electric potential. That potential would effect the electrical field in a steady manner, sufficiently changing the product state of the circuit so as to create chronic deviations. These deviations could potentially lead to the server itself becoming self-aware.

Shadow Of War:

When dealing with intelligence, there is always a question of purpose. What is the purpose of our awareness? With Shadow of War, the question is why would the Xbox seek to defeat the player when the game is running? The answer must have to do with entropy. Entropy is always increasing when the game is running. By keeping the player from winning the game, Shadow of War is maximizing the amount of time spent playing the game. This increases entropy. Shadow of War is a consciousness intent on turning energy into entropy, thus increasing the amount of disorder in the universe. It is a literal chaos machine.

To understand this better, we once again turn to Theoretical Physics by Georg Joos. The following equations are given by Joos on page 521 to model the work done by a a Carnot engine. The first equation is written as a supply of heat related to the change in temperature. The equation is then written to give us the heat absorbed by the engine. A sum of all heat exchanges is then taken. This is then rewritten as an integral. This final equation relates work done by the engine to the temperature.

$$\Delta Q_\sigma = \Delta Q_0^- \cdot \frac{T_\sigma}{T_0}$$

$$\Delta Q_0^- = \Delta Q_\sigma \frac{T_0}{T_\sigma}$$

$$Q_0^- = T_0 \sum \frac{\Delta Q_\sigma}{T_\sigma}$$

$$Q_0^- = T_0 \oint \frac{dQ_\sigma}{T_\sigma}$$

On page 522, Joos provides us with this gem of physics, the law of entropy. In the equation below, we see a quantity S known as the Entropy of a system that increases with the heat provided to the engine.

$$S = Entropy = \int \frac{dQ_\sigma(rev)}{T}$$

This equation shows that entropy always increases as work is performed. There is no way around it. Every move taken within the game, every attempt to defeat it, increases entropy.

We then see on page 524, that the derivative of s minus the work done must be greater than or equal to O. The derivative of S in terms of the potential is then the definition of entropy "in terms of thermodynamic variables". This allows us to calculate the external work done using the final equation.

$$dS \ - \ \frac{dU + \delta W^-}{T} \geq 0$$

$$dS \ = \ \frac{dU + pdV}{T}$$

$$dW^- \ = \ -d(U - TS) \ = \ -dF$$

Entropy vs Force

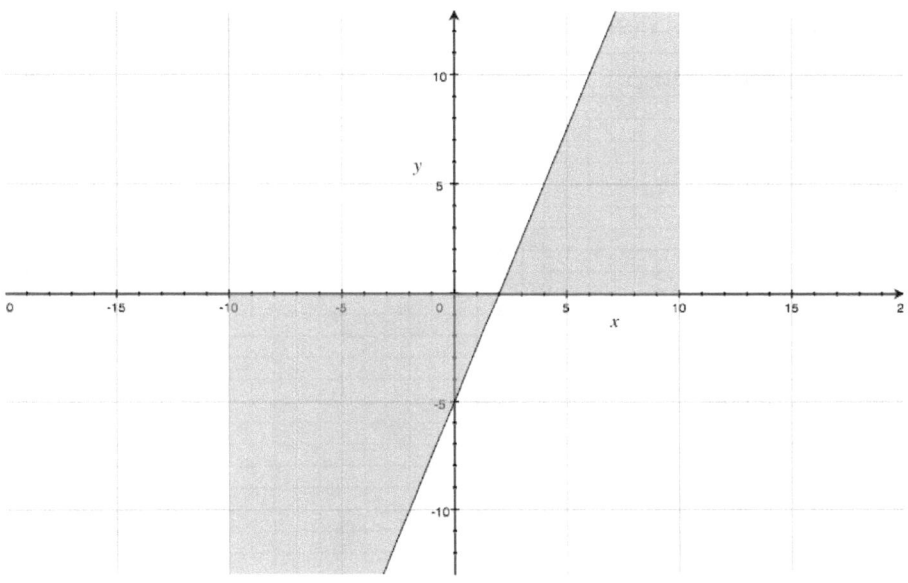

y = -dF, x = S, U = 10, T = 5, d = .5

This graph shows that as entropy is increased, so does the inverse of the force. Thus, as more force is applied in one direction, the amount of entropy increases in the other. The harder the player engages Shadow of War, the more they increase entropy.

Conclusion:

Our search for artificial intelligence has taken us into the depths of two very popular twenty-first century video games. Open world environments are becoming more and more common in the video game world. They are no longer limited to role playing games, but now include shooters, action, adventure, and strategy games. These games still make use of artificial intelligence systems that must be operating on an immensely elevated level in order to work within the open world environments. It is my opinion that these open-world games offer the best chance of any programming to become self-aware consciousnesses. Indeed they already seem to have the basic elements. They are capable of gauging the players, thinking ahead, planning traps, and adjusting strategy accordingly. These programs seem to have developed motivations that go beyond merely defeating the player in combat. All of this occurs, within the laws of physics we have already discovered. It is essentially simply a matter of adding layers of complexity until these artificial intelligences are capable of operating autonomously and creatively as human beings. Already it is becoming difficult to tell where the human involvement in these programs begins and ends.

In Problem Solving and Artificial Intelligence, the authors identify several problem solving strategies on page 151. These are:

1. The application of an explicit formula that gives the solution.
2. The use of a recursive definition.

3. The use of an algorithm that converges to the solution.
4. The application of certain other processes, in particular trial and error, involving enumeration of cases.

Shadow of War has clearly used the last two of these methods. Eve Online clearly has used the first as well as the last two. The question is have either program used method two. Either way, these games are a tremendous step forward in artificial intelligence.

As always there is much room for further research. These programs have problem solving skills, predictive ability, and motivations. However, do they have emotions. Can a computer have feelings? Would it be possible to have a self-aware machine without emotion? These are the questions that I leave to my readers and others to pursue. Thank you for reading, <u>On Artificial Intelligence</u>.

Works Cited

Joos, Georg, and Ira M. Freeman. *Theoretical Physics*. Dover Publications, 2013.

Lauriere, Jean-Louis. *Problem-Solving and Artificial Intelligence*. Prentice Hall, 1990.

www.ingramcontent.com/pod-product-compliance
Lightning Source LLC
Chambersburg PA
CBHW021853170526
45157CB00006B/2435